HOW TO SET UP YOUR AMAZON ECHO

A Complete Beginners To Pro Guide On How To Setup an Amazon Echo In 5 Minutes.

BY

DAVE BRUCE
Copyright©2018

COPYRIGHT

No part of this, publication may be reproduced, distributed, or transmitted in any form or by any means, including photocopy, recording or other electronic or mechanical methods, or by any information storage and retrieval system without the prior written permission of the publisher, except in a case of very brief quotations embodied in critical reviews and certain other noncommercial use per-mitted by copyright law.

Dave Bruce

TABLE OF CONTENT

CHAPTER 1 4
  INTRODUCTION 4
CHAPTER 2 6
  HOW TO SET UP AND CONFIGURE YOUR AMAZON ECHO 6
CHAPTER 3 9
  GETTING YOUR ECHO SET UP 9
CHAPTER 4 15
  SUMMONING ALEXA 15
CHAPTER 5 17
  THINGS YOU CAN DO WITH YOUR AMAZON ECHO 17
THE END 21

CHAPTER 1

INTRODUCTION

The Amazon Echo is one of the world most popular android tablets with over 3 million users who owns it, with the amazing features it has in it.

Apart from the popular use of Amazon Echo device as an E-reader, it can also perform other functions. The user interface is a friendly one and it has a lot of good stuff in it.

This guide will show you how you can setup your Amazon Echo, all you have to do to achieve and get acquainted with the Amazon Echo device is follow the steps as they are instructed here. Also you can as well perform wonders with the Amazon Echo features it has in it with proper meditations of each step provided here.

Dave Bruce

All over the world millions of people haven't been able to use and perform wonders with its features but dis book gives the breakdown of all solution to any problem you might encounter.
Thankfully each step is very easy and simple to follow, that even a beginner can master it in a few minutes.

CHAPTER 2

HOW TO SET UP AND CONFIGURE YOUR AMAZON ECHO

After purchasing your Amazon Echo, now we should take a look on how to set it up with some useful things your Echo can be task with.

What is The Meaning Of Amazon Echo?

Dave Bruce

The Echo is a voice controlled virtual assistant that enable you to control your home appliances, play music, get updated with news like sports, check weather, traffic, and lots more.
We have handful of different model that you can choose from, the new model $99 version that comes with a good speaker. The cheaper $50 Echo Dot, it comes with a laptop quality on the lower end. While on the higher end we have the $150 Echo Plus, having a built-in smart home hub and its speaker is slightly better than the $99 model. We have Echos with screens on them, like Echo Show and Echo Spot, it can be purchase.
Before setting the Echo up and use it, take a fast tour of the physical device and its buttons. Inside this guide we will show you how you can set up the latest $99 model.

There are three things on the Echo you can play with, the volume buttons, the action button (when pressed, calls Alexa without saying anything), and the microphone button (it enable the off and on listening feature). Previous generation Echos has a volume ring around its edge, rather than having the volume buttons.

All latest Echo devices have an audio-out port close to the power port at the bottom to enable you plug better and powerful speakers into them.

Now, let's proceed with how to set up your Echo

CHAPTER 3

GETTING YOUR ECHO SET UP

First step is, before removing Echo from its pack and Echo is plug in, grab the Amazon Alexa app for your tablet or phone (iOS/Android). Once app is downloaded, you should hold off on launching it right now.

After you have plug in your Amazon Echo, around the top the indicator ring will show blue then turn to a navigating orange color (as shown above).

After navigating to the orange color its signifies is ready to be configured. But in case you miss this step and the ring shows purple, hold the action button down until the ring shows orange again for about 5 to 6 seconds.

In the orange configuration mode once Echo is fully powered up, pull out your smart home then open the Wi-Fi settings. During the configuration process in order to set it up Echo requires you to connect directly to it.

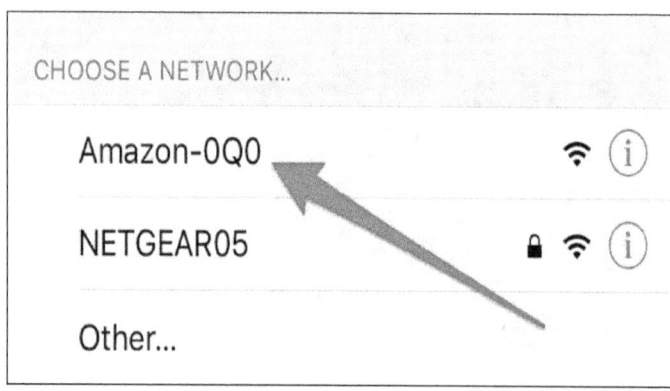

As shown above, you should choose the Echo's Wi-Fi connection, while trying to choose it will look like "Amazon-42N" or it may as well look like some other number and letter combo, and then connect to it. After that, to start the set up process you can launch the Alexa app and you should see this on your screen:

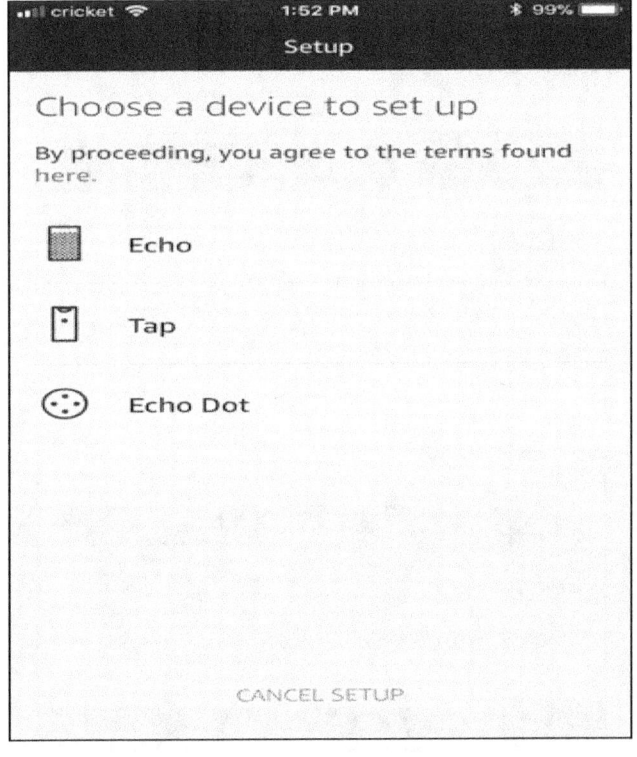

Choose the kind of Echo you are setting up and click on the menu icon and select settings, you should do this if the app didn't jump right into the configuration process. From there you can select either, you got an Echo as gift didn't personally purchase it, and then you should choose "Set up new device". Or, on your Amazon account you personally purchase the Echo, and then it should say "[Your Name]'s Echo, select it.

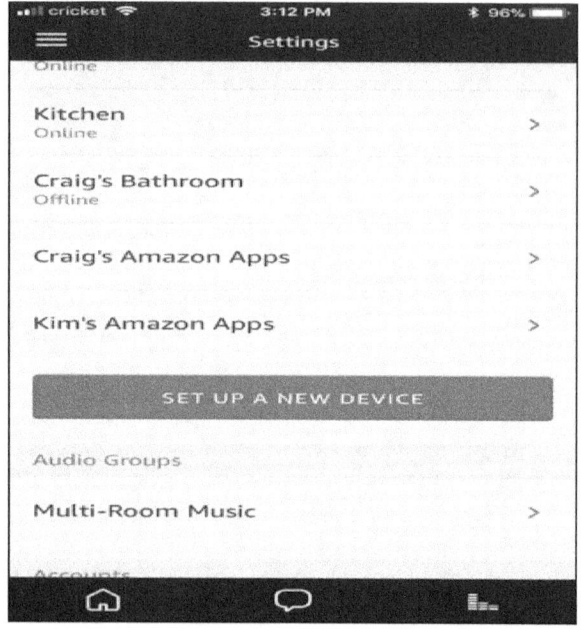

Dave Bruce

At the point where you are connected to the Echo with the app up and running, select your

Wi-Fi network from the list of networks the Echo can access after inputting your Amazon account login information's, agree to the Alexa user conditions (e.g. that you are not worried with your voice being sent to Amazon to be analyzed for commands and service improvements).

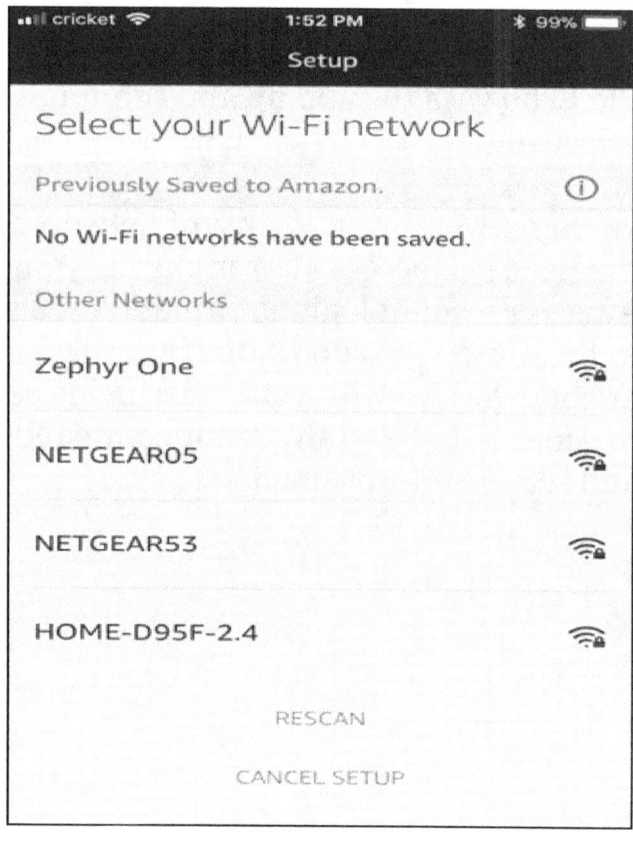

Once Echo is connected to home's Wi-Fi network, you are ready to start making use of it. Now let's take a look at how you can talk to Alexa and how you can customize your Echo experience.

CHAPTER 4

SUMMONING ALEXA

Anytime you say the word "Alexa", you should see the light ring atop the Echo light up blue, as the Echo is programmed to always respond to a "wake word", which is "Alexa" by default, (but if you wish you can change the wake word). You should then follow it up with a command or inquiry.

There is one command we want you to be aware of and have done pat, "Alexa, Stop". At some point you are going to ask Alexa to do something that leads to a lot of noise and you will definitely look for a way to stop it. It's also how music can be stop from playing, which includes alarms and timers.

CHAPTER 5

THINGS YOU CAN DO WITH YOUR AMAZON ECHO

Let's take a look at some of the amazing things you can perform with Alexa, now that you are acquainted with how to summon Alexa and put a stop to whatever things she might start.

Play Music

You may likely find yourself playing music through speakers as it is the least futuristic means of Echo unit. By means of default, your Amazon Echo is tapped into your Amazon Prime account and its enables you to purchase music through Amazon and library of free prime music. Linking your specify or Pandora is guarantee and to play music on multiple Echo at once all you need to do is just say words like:

Alexa, play (song) by (band)
Alexa, play (band)

Alexa, play (music genre)
Alexa, play (station) on (streaming service)
Alexa, what is playing now?
In order to make commands to Alexa to play any song of your choice if you have already created a playlist all you need to do is just call, "Alexa via playlist name" or you want Alexa to play a particular song just say, "Alexa play via song name".

Ask Alexa, Make Inquiries, and Get Conversions

You can ask Alexa questions make inquiries and get conversions on a broad range of topics and a direct response will be given to you by Alexa, but if Alexa is not programmed to the limit of your questions, she might give you a search results. To do these just ask:
Alexa, what is the news today?
Alexa, what time is it in (city/state/country)?
Alexa, what is 184 × 4?
Alexa, how bones are in the human body? (This and other straight conversion questions are easy for Alexa)

Alexa, what is the weather? (Add in [city name] if you want weather outside your zip code)

To-Do lists and Make shopping

If maybe you are productivity minded, the list functionality is handy and you can as well give commands to Alexa to create and add to your lists:

Alexa, read out my shopping list

Alexa, produce a new list

Alexa, add other (items) to my shopping list

Any item you add is stored in the app, as shown above, in Alexa app the list are transcribed for you. You can appeal it while running an errand.

Get Your Echo Experience Customize

If need be you may want to move to customizing your Echo experience by either downloading Alexa skills or adjusting some settings. To customize your settings open Alexa app, on the main menu, choose "settings", from there move down to "Account" section. You should find different options of things like music and sports update.

On Alexa skills with more functionality, access skills store by moving to main menu and select "Skills". Once you are there you will be able to download all sort of different skills, from ancient rain sounds to skills that will enable you have access to control all smart home devices. Thanks for reading my book and you can check out other books by same publisher.

HOW DO I SIDELOAD APPS INTO MY KINDLE FIRE - DAVE BRUCE

HOW TO SET UP YOUR NEW CHROMECAST – WILLIAMS ROGUE

HOW DO I INSTALL GOOGLE PLAY ON KINDLE FIRE – DAVE BRUCE

Dave Bruce

THE END

Dave Bruce

Dave Bruce

www.ingramcontent.com/pod-product-compliance
Lightning Source LLC
Chambersburg PA
CBHW071001220526
45471CB00007B/3130